Personal Information

Name:

Contact Information:

Education is not the learning of facts, but the training of the mind to think.

- Albert Einstein

Pages	Topic

Pages	Topic

Date: Topic: Page Number:

Subject: Instructions Course:

Cues / Questions

Notes

1

- Short Clear Sentences.
- List Key Ideas.
- Use Key Abbreviations And Paraphrase.
- Make Charts, Pictures, And Diagrams.

2

- Key Words And Cues To Jog Memory.
- Create Key Words And Phrases To Connect With The Notes And Jog Your Memory.

Summary

3

- Summarize 3-5 Key Points From Your Notes.
- Carry Over Your Points To The Adjacent Feynman Page.

Feynman Mastery: 3-5 Key Ideas Explained In Simple Language With Examples.

Instructions

Feynman Subject Mastery Method

Step 1: Choose 3-5 key concepts from your notes.

Step 2: Explain each concept as if you were explaining it to someone else. Use plain, simple language and examples.

Step 3: Review your explanation and identify any areas where you are uncertain or shaky. If you still don't understand, go back to the source material, your notes, or any examples you can find.

Step 4: If you've used lots of technical terms or complex language in your explanation, try re-writing it in simpler terms. Make sure anyone can understand what you're saying, even if they don't have the same knowledge you do.

Use another sheet of paper if necessary and attach it to this one when you finish.

Date: Topic: Page Number:

Subject: Course:

Cues / Questions

Notes

Summary

Feynman Mastery: 3-5 Key Ideas Explained In Simple Language With Examples.

Use another sheet of paper if necessary and attach it to this one when you finish.

Date: Topic: Page Number:

Subject: Course:

Cues / Questions

Notes

Summary

Feynman Mastery: 3-5 Key Ideas Explained In Simple Language With Examples.

Use another sheet of paper if necessary and attach it to this one when you finish.

Date: Topic: Page Number:

Subject: Course:

Cues / Questions

Notes

Summary

Feynman Mastery: 3-5 Key Ideas Explained In Simple Language With Examples.

Use another sheet of paper if necessary and attach it to this one when you finish.

Date: Topic: Page Number:

Subject: Course:

Cues / Questions

Notes

Summary

Feynman Mastery: 3-5 Key Ideas Explained In Simple Language With Examples.

Use another sheet of paper if necessary and attach it to this one when you finish.

Date: Topic: Page Number:

Subject: Course:

Cues / Questions

Notes

Summary

Feynman Mastery: 3-5 Key Ideas Explained In Simple Language With Examples.

Use another sheet of paper if necessary and attach it to this one when you finish.

Date: Topic: Page Number:

Subject: Course:

Cues / Questions

Notes

Summary

Feynman Mastery: 3-5 Key Ideas Explained In Simple Language With Examples.

Use another sheet of paper if necessary and attach it to this one when you finish.

Date:	Topic:	Page Number:
Subject:	Course:	

Cues / Questions

Notes

Summary

Feynman Mastery: 3-5 Key Ideas Explained In Simple Language With Examples.

Use another sheet of paper if necessary and attach it to this one when you finish.

Date: Topic: Page Number:

Subject: Course:

Cues / Questions

Notes

Summary

Feynman Mastery: 3-5 Key Ideas Explained In Simple Language With Examples.

Use another sheet of paper if necessary and attach it to this one when you finish.

Date: Topic: Page Number:

Subject: Course:

Cues / Questions

Notes

Summary

Feynman Mastery: 3-5 Key Ideas Explained In Simple Language With Examples.

Use another sheet of paper if necessary and attach it to this one when you finish.

Date: Topic: Page Number:

Subject: Course:

Cues / Questions

Notes

Summary

Feynman Mastery: 3-5 Key Ideas Explained In Simple Language With Examples.

Use another sheet of paper if necessary and attach it to this one when you finish.

Date: Topic: Page Number:

Subject: Course:

Cues / Questions

Notes

Summary

Feynman Mastery: 3-5 Key Ideas Explained In Simple Language With Examples.

Use another sheet of paper if necessary and attach it to this one when you finish.

Date: Topic: Page Number:

Subject: Course:

Cues / Questions

Notes

Summary

Feynman Mastery: 3-5 Key Ideas Explained In Simple Language With Examples.

Use another sheet of paper if necessary and attach it to this one when you finish.

Date: Topic: Page Number:

Subject: Course:

Cues / Questions

Notes

Summary

Feynman Mastery: 3-5 Key Ideas Explained In Simple Language With Examples.

Use another sheet of paper if necessary and attach it to this one when you finish.

Date: Topic: Page Number:

Subject: Course:

Cues / Questions

Notes

Summary

Feynman Mastery: 3-5 Key Ideas Explained In Simple Language With Examples.

Use another sheet of paper if necessary and attach it to this one when you finish.

Date: Topic: Page Number:

Subject: Course:

Cues / Questions

Notes

Summary

Feynman Mastery: 3-5 Key Ideas Explained In Simple Language With Examples.

Use another sheet of paper if necessary and attach it to this one when you finish.

Date: Topic: Page Number:

Subject: Course:

Cues / Questions

Notes

Summary

Feynman Mastery: 3-5 Key Ideas Explained In Simple Language With Examples.

Use another sheet of paper if necessary and attach it to this one when you finish.

Date: Topic: Page Number:

Subject: Course:

Cues / Questions

Notes

Summary

Feynman Mastery: 3-5 Key Ideas Explained In Simple Language With Examples.

Use another sheet of paper if necessary and attach it to this one when you finish.

Date: Topic: Page Number:

Subject: Course:

Cues / Questions

Notes

Summary

Feynman Mastery: 3-5 Key Ideas Explained In Simple Language With Examples.

Use another sheet of paper if necessary and attach it to this one when you finish.

Date: Topic: Page Number:

Subject: Course:

Cues / Questions

Notes

Summary

Feynman Mastery: 3-5 Key Ideas Explained In Simple Language With Examples.

Use another sheet of paper if necessary and attach it to this one when you finish.

Date:	Topic:	Page Number:
Subject:	Course:	

Cues / Questions

Notes

Summary

Feynman Mastery: 3-5 Key Ideas Explained In Simple Language With Examples.

Use another sheet of paper if necessary and attach it to this one when you finish.

Date: Topic: Page Number:

Subject: Course:

Cues / Questions

Notes

Summary

Feynman Mastery: 3-5 Key Ideas Explained In Simple Language With Examples.

Use another sheet of paper if necessary and attach it to this one when you finish.

Date: Topic: Page Number:

Subject: Course:

Cues / Questions

Notes

Summary

Feynman Mastery: 3-5 Key Ideas Explained In Simple Language With Examples.

Use another sheet of paper if necessary and attach it to this one when you finish.

Date: Topic: Page Number:

Subject: Course:

Cues / Questions

Notes

Summary

Feynman Mastery: 3-5 Key Ideas Explained In Simple Language With Examples.

Use another sheet of paper if necessary and attach it to this one when you finish.

Date: Topic: Page Number:

Subject: Course:

Cues / Questions

Notes

Summary

Feynman Mastery: 3-5 Key Ideas Explained In Simple Language With Examples.

Use another sheet of paper if necessary and attach it to this one when you finish.

Date:	Topic:	Page Number:
Subject:	Course:	

Cues / Questions

Notes

Summary

Feynman Mastery: 3-5 Key Ideas Explained In Simple Language With Examples.

Use another sheet of paper if necessary and attach it to this one when you finish.

Date: Topic: Page Number:

Subject: Course:

Cues / Questions

Notes

Summary

Feynman Mastery: 3-5 Key Ideas Explained In Simple Language With Examples.

Use another sheet of paper if necessary and attach it to this one when you finish.

Date: Topic: Page Number:

Subject: Course:

Cues / Questions

Notes

Summary

Feynman Mastery: 3-5 Key Ideas Explained In Simple Language With Examples.

Use another sheet of paper if necessary and attach it to this one when you finish.

Date: Topic: Page Number:

Subject: Course:

Cues / Questions

Notes

Summary

Feynman Mastery: 3-5 Key Ideas Explained In Simple Language With Examples.

Use another sheet of paper if necessary and attach it to this one when you finish.

Date: Topic: Page Number:

Subject: Course:

Cues / Questions

Notes

Summary

Feynman Mastery: 3-5 Key Ideas Explained In Simple Language With Examples.

Use another sheet of paper if necessary and attach it to this one when you finish.

Date: Topic: Page Number:

Subject: Course:

Cues / Questions

Notes

Summary

Feynman Mastery: 3-5 Key Ideas Explained In Simple Language With Examples.

Use another sheet of paper if necessary and attach it to this one when you finish.

Date: Topic: Page Number:

Subject: Course:

Cues / Questions

Notes

Summary

Feynman Mastery: 3-5 Key Ideas Explained In Simple Language With Examples.

Use another sheet of paper if necessary and attach it to this one when you finish.

Date: Topic: Page Number:

Subject: Course:

Cues / Questions

Notes

Summary

Feynman Mastery: 3-5 Key Ideas Explained In Simple Language With Examples.

Use another sheet of paper if necessary and attach it to this one when you finish.

Date: Topic: Page Number:

Subject: Course:

Cues / Questions

Notes

Summary

Feynman Mastery: 3-5 Key Ideas Explained In Simple Language With Examples.

Use another sheet of paper if necessary and attach it to this one when you finish.

Date: Topic: Page Number:

Subject: Course:

Cues / Questions

Notes

Summary

Feynman Mastery: 3-5 Key Ideas Explained In Simple Language With Examples.

Use another sheet of paper if necessary and attach it to this one when you finish.

Date: Topic: Page Number:

Subject: Course:

Cues / Questions

Notes

Summary

Feynman Mastery: 3-5 Key Ideas Explained In Simple Language With Examples.

Use another sheet of paper if necessary and attach it to this one when you finish.

Date: Topic: Page Number:

Subject: Course:

Cues / Questions

Notes

Summary

Feynman Mastery: 3-5 Key Ideas Explained In Simple Language With Examples.

Use another sheet of paper if necessary and attach it to this one when you finish.

Date: Topic: Page Number:

Subject: Course:

Cues / Questions

Notes

Summary

Feynman Mastery: 3-5 Key Ideas Explained In Simple Language With Examples.

Use another sheet of paper if necessary and attach it to this one when you finish.

Date: Topic: Page Number:

Subject: Course:

Cues / Questions

Notes

Summary

Feynman Mastery: 3-5 Key Ideas Explained In Simple Language With Examples.

Use another sheet of paper if necessary and attach it to this one when you finish.

Date: Topic: Page Number:

Subject: Course:

Cues / Questions

Notes

Summary

Feynman Mastery: 3-5 Key Ideas Explained In Simple Language With Examples.

Use another sheet of paper if necessary and attach it to this one when you finish.

Date: Topic: Page Number:

Subject: Course:

Cues / Questions

Notes

Summary

Feynman Mastery: 3-5 Key Ideas Explained In Simple Language With Examples.

Use another sheet of paper if necessary and attach it to this one when you finish.

Date:	Topic:	Page Number:
Subject:	Course:	

Cues / Questions

Notes

Summary

Feynman Mastery: 3-5 Key Ideas Explained In Simple Language With Examples.

Use another sheet of paper if necessary and attach it to this one when you finish.

Date: Topic: Page Number:

Subject: Course:

Cues / Questions

Notes

Summary

Feynman Mastery: 3-5 Key Ideas Explained In Simple Language With Examples.

Use another sheet of paper if necessary and attach it to this one when you finish.

Date: Topic: Page Number:

Subject: Course:

Cues / Questions

Notes

Summary

Feynman Mastery: 3-5 Key Ideas Explained In Simple Language With Examples.

Use another sheet of paper if necessary and attach it to this one when you finish.

Date: Topic: Page Number:

Subject: Course:

Cues / Questions

Notes

Summary

Feynman Mastery: 3-5 Key Ideas Explained In Simple Language With Examples.

Use another sheet of paper if necessary and attach it to this one when you finish.

Date: Topic: Page Number:

Subject: Course:

Cues / Questions

Notes

Summary

Feynman Mastery: 3-5 Key Ideas Explained In Simple Language With Examples.

Use another sheet of paper if necessary and attach it to this one when you finish.

Date: Topic: Page Number:

Subject: Course:

Cues / Questions

Notes

Summary

Feynman Mastery: 3-5 Key Ideas Explained In Simple Language With Examples.

Use another sheet of paper if necessary and attach it to this one when you finish.

Date: Topic: Page Number:

Subject: Course:

Cues / Questions

Notes

Summary

Feynman Mastery: 3-5 Key Ideas Explained In Simple Language With Examples.

Use another sheet of paper if necessary and attach it to this one when you finish.

Date: Topic: Page Number:

Subject: Course:

Cues / Questions

Notes

Summary

Feynman Mastery: 3-5 Key Ideas Explained In Simple Language With Examples.

Use another sheet of paper if necessary and attach it to this one when you finish.

Date: Topic: Page Number:

Subject: Course:

Cues / Questions

Notes

Summary

Feynman Mastery: 3-5 Key Ideas Explained In Simple Language With Examples.

Use another sheet of paper if necessary and attach it to this one when you finish.

Date: Topic: Page Number:

Subject: Course:

Cues / Questions

Notes

Summary

Feynman Mastery: 3-5 Key Ideas Explained In Simple Language With Examples.

Use another sheet of paper if necessary and attach it to this one when you finish.

Date: Topic: Page Number:

Subject: Course:

Cues / Questions

Notes

Summary

Feynman Mastery: 3-5 Key Ideas Explained In Simple Language With Examples.

Use another sheet of paper if necessary and attach it to this one when you finish.

Date: Topic: Page Number:

Subject: Course:

Cues / Questions

Notes

Summary

Feynman Mastery: 3-5 Key Ideas Explained In Simple Language With Examples.

Use another sheet of paper if necessary and attach it to this one when you finish.

Date: Topic: Page Number:

Subject: Course:

Cues / Questions

Notes

Summary

Feynman Mastery: 3-5 Key Ideas Explained In Simple Language With Examples.

Use another sheet of paper if necessary and attach it to this one when you finish.

Date: Topic: Page Number:

Subject: Course:

Cues / Questions

Notes

Summary

Feynman Mastery: 3-5 Key Ideas Explained In Simple Language With Examples.

Use another sheet of paper if necessary and attach it to this one when you finish.

Date: Topic: Page Number:

Subject: Course:

Cues / Questions

Notes

Summary

Feynman Mastery: 3-5 Key Ideas Explained In Simple Language With Examples.

Use another sheet of paper if necessary and attach it to this one when you finish.

Science-Based Study

Self-testing

Make questions about the material and then quiz yourself. It's confirmed by research as a great way to learn and remember.

- **Make questions from your notes** as you go along studying and put the questions on the front of flashcards.
- **Write the answers on the back** of the flashcards …
- … and then you practice answering the questions.

Spaced Practice

Begin at least 4 weeks before exam date. Study at spaced intervals to connect ideas, expand knowledge, and recall more easily.

Follow this schedule:

- **Day 1:** Take notes and make study flashcards with questions on front and answers on back.
- **Day 2:** Revisit and review the notes and quiz yourself with flashcards.
- **Day 3:** Again, review the notes and quiz yourself with flashcards.
- **After one week:** Once again review the notes and quiz yourself with flashcards.
- **After two weeks:** Finally, review the notes and quiz yourself with flashcards.

Elaborative Learning.

It's a method of asking questions about your subject that makes you to think about it in many different ways.

- **Make a list of the things you need to learn** from your class materials. Ask yourself questions about how these ideas work and why. As you ask yourself questions, search your study materials and look for the answers to the questions.
- **Continue to elaborate on the ideas you are learning.** Make connections between the different parts of your subject and explain how they work together. A good way to do this is to compare two ideas and think of how they are similar and how they are different.
- **Think about how the different ideas you are studying fit in your own life experiences** and memories. Also, as you go through your day, pay attention to things happening around you and make connections with things you are learning in class.

Made in the USA
Monee, IL
02 August 2022

10737303R00070